IMAGES
of America

BREWING IN MONTANA

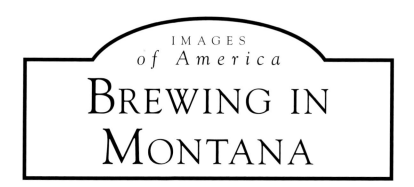

IMAGES
of America

BREWING IN MONTANA

Ryan Newhouse

ARCADIA
PUBLISHING

Published by Arcadia Publishing
Charleston, South Carolina

Printed in the United States of America

Library of Congress Control Number: 2015959856

For all general information, please contact Arcadia Publishing:
Telephone 843-853-2070
Fax 843-853-0044
E-mail sales@arcadiapublishing.com
For customer service and orders:
Toll-Free 1-888-313-2665

Visit us on the Internet at www.arcadiapublishing.com

.

To Steve Lozar, who holds the keys to Montana's brewing history.
I cannot wait to read your book.

CONTENTS

ACKNOWLEDGMENTS

Today's Montana breweries, all 90-plus of them, are in their own ways continuing Montana's 159-year history of brewing beer for its men and women. If they did not strive to make great beer today, I fear we would not care as much about the brewers who came before.

I especially want to thank Steve Lozar, founder and curator of Lozar's Montana Brewery Museum. So many images appear in this book courtesy of Lozar's Montana Brewery Museum. Beyond imagery, it is Steve's tireless and keen research that has allowed these pages to be filled with stories and facts. You will hear his talks and articles echoed here. His work has led to the rediscovery of many Montana breweries that would have been forever lost to the beer drinkers and history buffs like me. Steve has my immeasurable gratitude for opening up his museum to me and making it one of the best experiences I have had in Montana.

There are many others I owe thanks to, including my editor at Arcadia Publishing, Stacia Bannerman, who prodded me every step of the way.

I also want to thank Gary Flynn of BreweryGems.com. Not only has Gary provided invaluable images and historical research to this work but also to my first book, *Montana Beer: A Guide to Breweries in Big Sky Country*. Please check out his website and be as impressed as I was with his work.

Of course, I appreciate my family, friends, and all others with whom I have shared beers and by whom I was supported in all my Montana beer endeavors. Notably, I thank Kate Bernot for being my concert mate and the most inspiring of all beer writers. And I thank my partner Kristin Eckholm, with whom I want to explore every inch of Montana. They are welcome and steady influences.

INTRODUCTION

The brewing history of Montana is a mirror of the history of the territory and state. Seemingly every mining camp, cow town, logging center, and city had its own local breweries. Author Ryan Newhouse presents in this, his second book on Montana's brewing legacy, a graphic illustration of over 150 years of brewing in the state he now calls home. His words and the photographs bring to life the colorful history of one of Montana's earliest and most vibrant industries.

Montana brewers of the past created their beers to satisfy the tastes of their local citizenry. With few exceptions, the owners and operators of the community brew houses were German and Irish immigrants. They brought from their homelands the styles of brews that were popular where they had learned their art and craft. They tended to land in Montana communities that had high numbers of people with their own ethnic background and tastes. These new Montana brewers enthusiastically embraced America while still maintaining the taste comforts of their places of birth. Newhouse has traveled Montana, experiencing the diversity of the neighborhoods where breweries once flourished. His understanding of today's Montana craft-brewing landscape has sharpened his ability to recognize the economic and cultural impact the industry has on those same communities that traditionally enjoyed their hometown beer. He deftly weaves that connection of the significance of the past throughout this book.

Brewers in Montana have always been faced with challenges that have affected their ability to produce a quality product and still remain economically viable. Those challenges included keeping a consistent supply of brewing materials, reliable and affordable transportation, effective marketing, a dependable trained workforce, and brews that were favored over regional and national brands. Territorial, state, and federal laws governing the beer business as well as Prohibition have had profound effects on the industry. Montana brewers have a long history of creatively addressing those effects.

Beer advocating advertising has been the most powerful tool at the disposal of Montana's brewing industry. Ryan Newhouse has artfully displayed the clever ways the consuming public was made aware of the laurels of Montana's local libations. Traditionally, the state's breweries used newspapers, signs, matchbooks, combs, brushes, silverware, belts, billboards, makeup compacts, records, glasses, sewing kits, first-aid kits, pens, pencils, whistles, trays, thermometers, bingo cards, fans, sports pools, shotgun shells, lighters, banners, bottles, cans, and sports teams. All these and much more were employed to get their brewery name out in front of the daily consuming public. Newhouse has displayed some of the examples of this promotional "breweriana" well.

Much of what the author characterizes in this book can be represented in this excerpt from an 1897 Montana newspaper article:

> Having here a local brewery, making a pure and a strictly unadulterated article, has been no small factor in removing prejudice against this great American beverage. The Dillon brewery is the means of thousands of dollars annually being kept in circulation

in Beaverhead county which otherwise would go to barley growers, manufacturers and workingmen of other states. The home brew is free from poisonous agencies and this one fact alone causes those careful of health to call for it. It is a distinct local institution, paying its quota of local taxation; contributing its share toward every local public and semi-public enterprise; affords a good home market for Beaverhead barley; gives labor to Dillon laborers: disseminates more or less money in all local trades and institutions and its power for good in all this can yet be increased three-fold if those who drink beer call for the home made product. Give Dillon beer a fair show. Quit enriching outside communities and stand by local enterprise.

The health issues addressed by Montana's brewers ranged from the necessity of breastfeeding mothers to drink plenty of their "liquid bread" for their children's nourishment to this public warning: "TYPHOID Lurks in Unclean Water, but not in our beer as it is boiled for hours and no germ life can live in it." Brewers in the patent-medicine era claimed that for thin men, their beer should be used "not once but three times a day. Surer than medicine and twice as pleasant to taste." They gave advice to "the Woman of the House, upon her often falls a heavy burden, the daily routine of housework, the care of children, the shopping, the social duties. Small wonder that she often sustains a 'breakdown' and must receive medical assistance. Such a result may be avoided by the moderate use of American [Great Falls] Beer." For the working man, local beer provides "one of the greatest enjoyments of a life of toil and the wholesome nourishment it contains lends strength and vigor to his system." As Newhouse labored over the creation of this book, he need not "Worry and Overwork" because he undoubtedly knew that both maladies "Produce the same effect—nervous depression." Certainly he took the Montana brewers' advice in 1908, who counseled, "Beer is a splendid restorative and invigorator. The hops produce a quieting and tonic effect that are a boon to shattered nerves, and induce refreshing sleep. The nutritive element of the barley assists in rebuilding wasted tissues."

Union labor has helped define Montana's brewing past. The early days of transient breweries in the silver- and gold-rich gulches of Montana saw very independent operations. As the territory's population increased and the demand for more beer became greater, breweries enlarged to slake the thirst of the state's residents. With growth came labor unions. Unionized breweries became the norm as Montana reached statehood in 1889. The slogan in Montana breweries became, "A Home Product, Strictly Union Made." The Montana Brewers Association became the statewide organization of its breweries to negotiate contracts with the unions. While the negotiations were sometimes contentious, the eventual settlements brought the public much joy and conviviality.

As readers enjoy Ryan Newhouse's book, I suggest you do so with a glass of Montana-made beer. For like a fine local brew, it promises to "furnish animation, sharpen the wits, promote sociability and make conversation flow."

I recommend Ryan's book with a hearty prost, slainte, and *na zdravje*!

—Steve "Bubs" Lozar
Montana Brewery Historian

One

MISSOULA

Missoula is known as the Garden City, and in 1895, the Garden City Brewery was founded by Joseph Steiger, Joseph Redle, and Paul Gerber. Since there was little reliable access to transportation suitable for beer, most of the Missoula-made beer was consumed there too.

The most famed beer to come out of Missoula is by far Highlander, which debuted in 1910. However, before Highlander could be used as a beer name, the owners of Garden City Brewery had to write a letter and ask permission from a New York baseball team with the same name. The team gave its approval, as it was preparing to change its name to what it is still called today, the New York Yankees.

Garden City Brewery enjoyed mild success until Prohibition, but it needed new owners and a new name in 1934, becoming the Missoula Brewing Company. (Joseph Wagner had opened a previous Missoula Brewing Company in 1890 at 109 West Front Street, but it did not last).

Over the next 30 years, Highlander still reigned supreme. In 1944, Emil Sick bought the brewery and operated the business until the economic slump of the 1960s. In 1964, Missoula Brewing Company was razed, making way for Interstate 90.

Highlander was resurrected in 2008 by Bob Lukes, who also brought back Missoula Brewing Company, which opened in July 2015.

The Garden City Brewery opened two years after the University of Montana–Missoula began classes. Its operating capacity was 15,000 barrels annually in 1900. This photograph was taken in 1902. Prior to creating Highlander, the brewery was best known for brewing Extra Pole Bohemian beer. (Courtesy of Bob Lukes, Missoula Brewing Company.)

Some of the men who worked at Garden City in the early 20th century included Joe Steiger (back row, left), Henry Emmerich (back row, fourth from right), and Joe Riddle (back row, right). High school and college-age workers were required to unload 100-pound sacks of barley from railcars before moving on to work the heavier bottle-washing machinery. About one in five applicants did not last the first day. (Courtesy of the HMFM Irene Dolan Collection 1979.32.2.)

Pictured are some of the oldest surviving Highlander bottles. The three on the left date from between 1910 and 1918, when brewed by Garden City Brewery. The one on the right is post-1934, after the brewery was renamed Missoula Brewing Company. (Courtesy of Lozar's Montana Brewery Museum.)

Highlander was also available in cans. This cone-top can is harder for collectors to find today. (Courtesy of Lozar's Montana Brewery Museum.)

Henry Emmerich (right) stands by brewing equipment inside the Garden City Brewery. (Courtesy of the HMFM Irene Dolan Collection 1979.32.3.)

In this later image of the brewery, the smokestack was rebuilt and labeled "Highlander." Horse-drawn buggies were replaced with automobiles. (Courtesy of Bob Lukes, Missoula Brewing Company.)

The Louvre Saloon inside the old Hotel Florence likely poured Garden City Brewery beer, given that pictured here around 1907 at right is Henry Emmerich, who worked at the brewery. Post-Prohibition, the lines between bars and breweries had to be more clear; no brewery could also own a bar. (Courtesy of the HMFM Irene Dolan Collection 1979.32.7.)

In the early 1900s, a glass of Garden City Brewery beer cost 5¢ at the 4 Mile Exchange in the Cold Springs area of Missoula. Henry Emmerich owned the bar. (Courtesy of the HMFM Irene Dolan Collection 1979.32.9.)

A worker stands inside the Garden City Brewery. The owners tried to sell in 1922, but no one was willing to invest in a brewery during Prohibition. They survived by offering near beer and other nonalcoholic beverages. (Courtesy of the HMFM Irene Dolan Collection 1979.32.20.)

In 1934, Highlander started being brewed under new ownership. Missoula Brewing Company would see the Highlander brand through to the end. Employees would commonly get beer as part of their wages, and they had their own large pails, which they filled several times a day at a tapped keg of Highlander. (Courtesy of Lozar's Montana Brewery Museum.)

Like days of old when minstrels and bards would share songs for beer and lodging, the Missoula Brewing Company made singing part of its marketing and branding efforts. Here is "At a 'Shindig'" with the tagline at bottom, "Keep a Case in Your Home." (Courtesy of Lozar's Montana Brewery Museum.)

SONGS OF OLD BUTTE—*By Joseph H. Duffy*

AT A "SHINDIG"

Folks do not dance like we used to dance
    At a "shindig";
In the days gone by, one took a chance
    At a "shindig."
But times have changed from the days "way back,"
When we danced all night at a miner's "shack"—
Of fun galore, there was little lack
    At a "shindig."

Old Brown "th' fiddler," he'd drop in
    At a "shindig";
And he'd bring along his violin
    To the "shindig."
And he'd fiddle long and he'd fiddle loud—
His shoulders bent and his old head bowed;
A wandering minstrel in the crowd
    At a "shindig."

The girls were young and the girls were fair
    At a "shindig."
The boys were a brand of "divil-may-care,"
    At a "shindig."
And the wit was fast and the wit was free
And the long night rocked with gales of glee;
Dawn came to quick for you and me
    At a "shindig."

Old Brown would fiddle jigs and reels
    At a "shindig."
The floors would creak with dancing heels
    At a "shindig."
An amber brew refreshed us then,
It came in kegs of "five" and "ten,"
For flasks on hips, there was no "yen,"
    At a "shindig."

And lovers oft would bill and coo
    At a "shindig."
They'd do their "petting" in full view
    At a "shindig."
And many, who were lovers then,
Have girls grown up—their boys are men,
Yet oft they yearn to be again
    at a "shindig."

Old mem'ries sometimes takes us back
    To a "shindig";
To the revels of a humble "shack"
    And a "shindig."
Oh dear old nights of happiness
That could not last—yet let's confess
The joy we'd feel in the light caress
    Of a "shindig."

HIGHLANDER

PALE EXPORT

BEER

BOTH PHONES   SOLD EVERYWHERE

The actual style of Highlander beer changed several times during its long life. Here it is advertised as a pale export. After being reintroduced in 2008, it was a Scottish ale. This advertisement, sticking with its musical connections, was featured on the back of an Elks program from 1911, featuring the All Star Minstrels. (Courtesy of Lozar's Montana Brewery Museum.)

"MY MONTANA"

Montana. Montana.
Where things are just about the way you like things to be.
Montana. Montana.
Montana is the only place for me.
      There is copper, there is oil
      There are silver-studded boulders
      Golden Sun on Golden Grain
      There are mountains that are peekin'
      Over one another's shoulders
      Lookin' at the great, Great Plain.
Montana. Montana.
Where things are just about the way you like things to be.
Montana. Montana.
Montana is the only place for me.
      There's a hundred fifty thousand miles
      Of free and easy livin'
      Air so fresh, it's sweeter than perfume.
      And everywhere you look you see
      The treasures we've been given —
      Lots of friends and lots and lots of room.
Montana. Montana.
Montana is the only place for me.

HIGHLANDER BREWING CO., SEATTLE, WASH., U.S.A.

Likely after Emil Sick purchased the brewery in 1944 and began moving some of Highlander's production farther west, near his Rainier Brewing Company, the brewery produced this song, "My Montana," and released it on LP. (Courtesy of Lozar's Montana Brewery Museum.)

As it still is today for many Montanans, beer and the outdoors paired well for drinkers of Highlander beer. This Outing Guide encouraged recreationists to clean up after themselves and ensure all campfires are put out completely. While the pamphlet is not dated, the use of the slogan "only you can prevent forest fires" was not created until 1947, three years after the introduction of Smokey the Bear. (Both, courtesy of Lozar's Montana Brewery Museum.)

U. S. Forest Service photo.

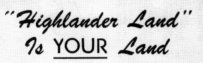

"Highlander Land" Is YOUR Land

. . . KEEP IT CLEAN
. . . KEEP IT GREEN

Western Montana's vast outdoors offers many miles of first class highways that follow fishing streams, meander over mountain passes and along lake shores.

As you travel in this "Land of the Shining Mountains" you can't help but feel the warmth and hospitality of a people who are proud of the area in which they live and take sincere delight in welcoming the stranger and acquainting him with the unspoiled country he is visiting.

All that is asked is that the visitor, after enjoying the facilities of campsites or roadside parks, observe the golden rule of leaving a clean camp to be enjoyed by the next traveler. Be sure that all camp fires are thoroughly extinguished, and, when traveling, don't be a flipper . . . use your ashtray and keep a litter bag in your car.

Only YOU Can Prevent Forest Fires

If you plan on going "off the beaten track" into forested areas during the summer forest fire season you must carry in your car a serviceable shovel, axe and bucket.

This part of Montana, in which we hope you will spend more than a little time, has been blessed with an abundance of natural beauty . . . thanks for helping us keep it that way.

Designed by Copi-Art, Anaconda, Mont.

McKEE ⬥ PRINT.

17

For advertising, beer brands were often attached to everyday items, like this book of matches. (Courtesy of Lozar's Montana Brewery Museum.)

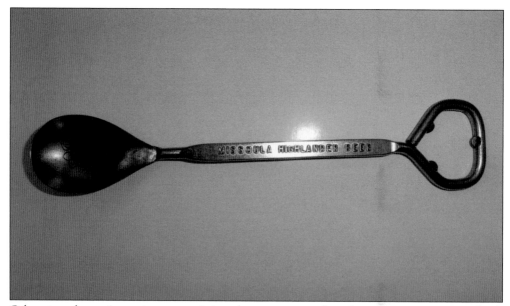

Other times, breweriana—items containing a brewery name or brand name—was more specialized, like this bottle opener–spoon. One could imagine it useful for the father or mother who sat down at the table, opened a bottle of Highlander, and used the spoon to feed junior. (Courtesy of Lozar's Montana Brewery Museum.)

Pictured here is one of the sleekest renditions of the Highlander beer can; this design was produced out of Seattle, Washington, in the 1960s through Sick's Rainier Brewing Company. (Courtesy of Lozar's Montana Brewery Museum.)

A late-style Highlander label carried Emil Sick's "6" mark, which appeared on other brands he owned at the time. It reads, "Symbol of Quality & Symbolic Pronunciation." (Courtesy of Bob Lukes, Missoula Brewing Company.)

Pictured here is a standard-issue Highlander bottle opener. These were available in a multitude of colors and were likely once found in every adult pocket in Missoula. (Courtesy of Lozar's Montana Brewery Museum.)

This Highlander "ghost sign" was found on a building in Chinook, Montana. (Courtesy of Jimmy Emerson, DVM.)

In Butte, Highlander Beer was dispensed by Bertoglio McTaggert distributors. The Bertoglio name is seen faintly at the top of the building. Located in the warehouse district, beer and many other items could come in on any of the five trains that arrived daily. (Courtesy of Larry Myhre.)

In the mid-1950s, under the guidance of Emil Sick, the Highlander brand adopted its famed tartan style. The brewery was sold at public auction on October 20, 1964, and the buildings razed to accommodate the building of Interstate 90. (Courtesy of Lozar's Montana Brewery Museum.)

Though not produced in great quantities, the Highlander brand had its relaxed style, too. Perhaps these were not only decorative, but like lids on steins, could keep the bugs at bay between sips of beer. (Courtesy of Lozar's Montana Brewery Museum.)

Two

BUTTE

After the end of the Civil War, miners flocked to Butte. As many of them came from Central and Eastern Europe, they held a strong fondness for their native beers, including lagers, bocks, and pilsners. More enriching, they also brought their favorite recipes for these beers, including porters, stouts, and cream ales.

The breweries that made Butte an epicenter for beer in the 19th and 20th centuries included Centennial Brewing Company, Butte Brewing Company, Tivoli Brewing Company, Silver Bow Brewing, and Olympia Brewing (not directly related to the Washington brewery). Of the bunch, Centennial Brewing Company made the most beer, but Butte Brewing Company lasted the longest. Each had a role to play in the culture-rich and diverse mining city.

While each brewery in Butte needed to build close to a water source (like Silver Bow Brewery on Silver Bow Creek), they were also built in proximity to ethnic enclaves. As Steve Lozar writes in his detailed article, "1,000,000 Glasses a Day: Butte's Brewing History," "Often a beer drinker's occupation, socioeconomic status, place of residence and fraternal organizations could be generally identified just by noticing which brand of beer they hoisted to their lips. Butte also took great pride in its unionism, and all the breweries were union shops."

Here is a common black notebook that would be carried by a brewmaster. This one was used by a brewer at Butte Brewing Company. (Courtesy of Lozar's Montana Brewery Museum.)

In addition to beer recipes, this notebook has a recipe for "vaginal itch," written out presumably by the brewmaster himself. The note at bottom reads, "1/3 of [American] women suffer external vaginal itching." (Courtesy of Lozar's Montana Brewery Museum.)

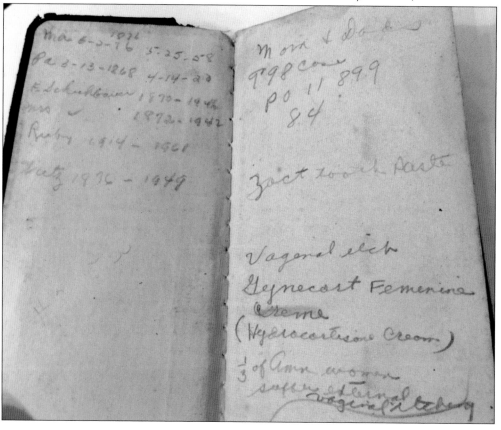

Butte Special was the best-known beer from Butte Brewing Company. In many advertisements, it was dubbed "Montana's Finest Beer." (Courtesy of Lozar's Montana Brewery Museum.)

Butte Brewing Company's founders were said to have "brought the luck and pluck of the Irish." The bock beer style was a favorite of many, including Irish miners. (Courtesy of Lozar's Montana Brewery Museum.)

German brewer Henry Muntzer established the Butte Brewing Company in 1885. Within 12 years, he had grown it from a five-barrel-a-day operation to 50 barrels a day. (Courtesy of the Butte Digital Image Project at Montana Memory Project.)

Pictured here is the inside of Butte Brewing Company; clockwise from the bottom are the filling room, fermenting room, brew kettle, malt room, mash tun, cellar, and bottling house. The Butte Brewery was near Finntown, and it played a role in the Finns', Swedes', and Norwegians' annual festivities, even hosting impromptu lutefisk-eating and beer-drinking competitions. Also common were street dancing and bare-fist boxing matches. (Courtesy of the Butte Digital Image Project at Montana Memory Project.)

Butte Brewing Company marketed its brand through household items, like this ashtray and collection of pens. Beer was part of daily family life in Butte. (Both, courtesy of Lozar's Montana Brewery Museum.)

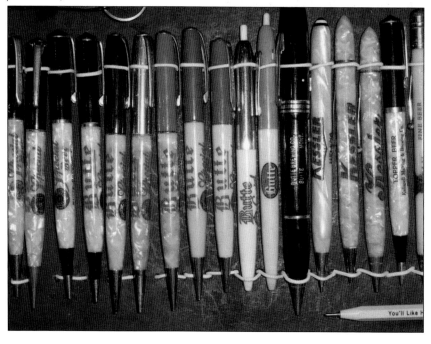

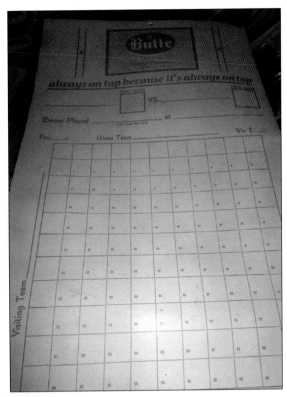

Sports betting at bars often took the shape of a box, like this one featuring Butte Brewing Company's advertising. Sometimes the back could serve as an impromptu event poster. This particular board was hung up at a Masonic bar, the Fez. It advertised the Clay-Liston fight (Cassius Clay had not yet changed his name to Muhammad Ali). *Sports Illustrated* magazine named this the fourth-greatest sporting event in the 20th century. Sonny Liston gave up in the seventh round. (Both, courtesy of Lozar's Montana Brewery Museum.)

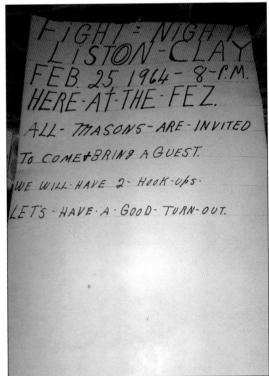

Several ghost signs remain in Butte today. After Prohibition, the Butte Brewing Company was the only local brewery to reopen. (Above, courtesy of Jon Wick; below, courtesy of Larry Myhre.)

This is Centennial Brewing Company's letterhead from 1907. When producing 1.3 million gallons of beer annually, the brewery used five million pounds of Montana barley and 85,000 pounds of hops, and the annual payroll for its unionized workforce of 100 totaled $144,000. (Courtesy of BreweryGems.com.)

In the early days, as pictured here around 1880, Centennial Brewing Company could only produce 500 barrels a year but later built a three-story malt house and steam-powered brew house, increasing its capacity to 6,000 barrels annually. After these improvements, the brewery sold its beer for $3 per dozen. At its peak, the brewery produced 40,000 barrels of beer a year. Centennial coined the slogan "A Million Glasses a Day—Someone Must Like It!" in 1905. (Courtesy of Lozar's Montana Brewery Museum.)

As the brewery grew, Centennial Brewing Company learned how to market not only to men, but women as well, as one 20th century advertisement read, "Health and beauty . . . are both won by drinking our wholesome and strengthening Export Beer bottled expressly for family use. It will resuscitate the jaded spirits to life and give tone and vigor that brings the beauty of health to the weak and run down system. For tired or debilitated women there is nothing that gives such good effects as our beer. It is a delightful and pleasant drink." (Courtesy of Lozar's Montana Brewery Museum.)

Olympia Brewing Company lasted 12 years, from 1899 to 1911. It is believed that Henry Mueller, president of the Centennial Brewing Company, was part of the brewery, along with Louis Best, of the Best family of Milwaukee—owners of Pabst Brewing. Mueller and Best also established the Billings Brewing Company at about the same time. This photograph features some of its workers in 1901, right before the brewery launched an expansion project. In the center is the brewmaster, John Weidenfeller, with the vest. (Courtesy of BreweryGems.com.)

One of the workers from the above photograph (on the far left of group) is pictured next to newly installed refrigeration equipment after the brewery expanded in 1902, which doubled output to 50 barrels a day. (Courtesy of BreweryGems.com.)

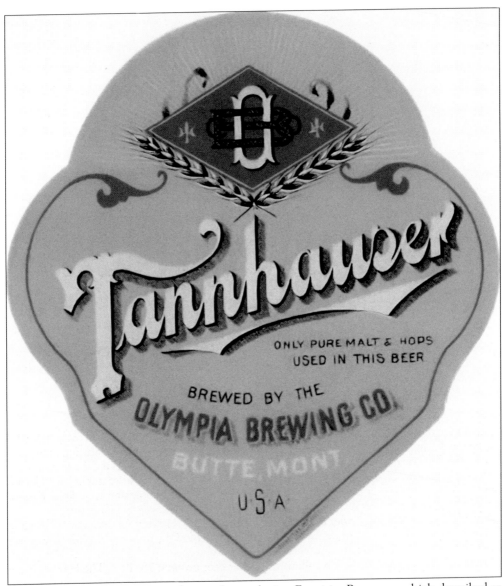

Although Olympia Brewing was known for marketing Exquisite Brew, was which described as an Andechser lager, a style of lager made at Kloster Andechs, a monastery southwest of Munich, Tannhauser was introduced by Gustav Hodel, a native of Baden, Germany, and the brewery's initial brewmaster. Hodel transferred from the Billings Brewing Company, which was also owned by Henry Mueller. Hodel had worked at the Silver Bow and Centennial Breweries before earning his master brewer's credentials in early 1900. He assumed his German friends would know Tannhauser, as it was the name of a famous opera written in 1845 by German composer Richard Wagner. (Courtesy of BreweryGems.com.)

Three

HELENA

In the tomes of Montana brewery history, Helena is essentially synonymous with Kessler Brewing Company. Its tenure from 1865 to 1958 made it one of Montana's longest running breweries. The brewery was started by Charles Beeher, the same entrepreneur who started the first brewery in Virginia City (what became the H.S. Gilbert Brewery). Three years after starting his Helena brewery, which he called the Ten Mile Brewery, he sold it to Luxembourg-born Nickolas Kessler, a man who dabbled in everything from gold prospecting to trade and manufacturing.

By 1891, Nickolas's son Charles Kessler began working at the brewery. Charles studied brewing in Chicago at the Henius Institute. By 1901, he was president of the brewery. In 1902, he founded the Montana State Brewers Association. He fought Prohibition hard, though the US collector of internal revenue destroyed 588 gallons of Kessler beer on June 28, 1919. The brewery came back after Prohibition, but it struggled for 25 years. In its last year of production, the brewery was capable of making 40,000 barrels annually but sold only 4,000.

In 1984, the name of Kessler beer was revived by Montana Beverages Ltd. However, it was not tied to the Kessler family or the brewery. The new Kessler Brewery operated in downtown Helena until 2000.

Breweries helping other breweries was as common in the early 20th century as it is today. Here is a bill of sale for malt from Kessler Brewing Company to the American Brewing Company in Great Falls, Montana. (Courtesy of Lozar's Montana Brewery Museum.)

Pictured here is an early example of the classic six-pack that made it convenient for women (the target demographic) to bring beer home from the grocery. (Courtesy of Lozar's Montana Brewery Museum.)

Kessler Brewing Company's Lorelei beer was well known. Here it is prominently featured on a beer-serving tray, commonly used by the Montana bars that served Kessler beer. (Courtesy of Lozar's Montana Brewery Museum.)

Continuing the tradition of advertising through common household items, here is a shoe-polishing brush with the Kessler name. (Courtesy of Lozar's Montana Brewery Museum.)

As was the case with milk, local breweries would deliver beer to the front door of homes. Kessler beer would likely arrive in a box like this one. (Courtesy of Lozar's Montana Brewery Museum.)

Kessler Brewing Company produced "The Story of Kessler Brewery," which outlined its history and explained how the brewery made beer. (Courtesy of Lozar's Montana Brewery Museum.)

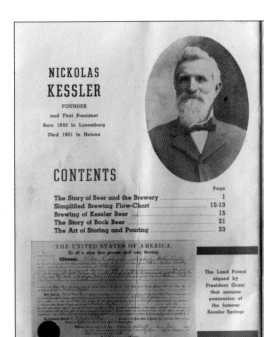

NICKOLAS KESSLER

FOUNDER

and First President

Born 1833 in Luxemburg

Died 1901 in Helena

CONTENTS

THE UNITED STATES OF AMERICA.

The Land Patent signed by President Grant that assures possession of the famous Kessler Springs.

the Story

Thirsty Montanans call it a cooling, satisfying, refreshing beverage. The hostess welcomes it as an aid to gracious entertaining. The cook uses it to add zest to otherwise mundane dishes.

From it, the rancher obtains revenue, the worker employment, the grocer and tavern keeper an important item for sale. Its production provides work for the barrel maker, the glass and can manufacturer, the trucker, the salesman, the hotel keeper and restaurant operator. To the Government, it is an important source of taxes. Because of it, the wheels of our economy turn faster, and, we hope, a little more pleasantly. Increasingly, thousands of Montanans and millions of Americans consider it the drink of sociability; a wholesome, friendly beverage of moderation.

It is beer, an integral part of the world for thousands of years. Although it is accepted and enjoyed like other good, natural products such as butter and milk few people realize the skill and science that is required to manufacture a quality product.

This booklet endeavors to tell a little about beer by telling you of Montana's oldest manufacturing establishment . . . the Kessler Brewing Company, started in 1865, more than 87 years ago.

First, let us look backwards into by-gone ages. Beer is as old as man and as world-wide. It is known that

1

According to "The Story of Kessler Brewery," "Thirsty Montanans call it a cooling, satisfying, refreshing beverage. The hostess welcomes it as an aid to gracious entertaining. The cook uses it to add zest to otherwise mundane dishes." (Courtesy of Lozar's Montana Brewery Museum.)

personal than today's, perhaps. Indians and road-agents harrassed the country. Transportation was slow and precarious. Many wagon trains carrying loads of barley malt from Utah 500 miles away, never arrived at all. Obtaining suitable hops was a major problem. The spirit of "make-do," of improvization pervaded throughout the mining camp. When brewing was held up because the order of hops had not arrived, native herbs such as wild grape roots and even the tender needles of the spruce were used. Because there was no hardwood, many of the barrels containing Kessler beer were cut-down and coopered whisky barrels. Despite all of these obstacles, the brewery grew. Some new equipment was added, new buildings erected, but it was not until 1886 that revolutionary advancement was made.

In that year, Kessler installed what was probably the first ice-making machine in any establishment between Chicago and the Pacific Coast. Because American manufacturers had not developed a satisfactory unit, a 15-ton carbonic acid gas refrigerating machine was purchased from the Krupp Machine Works, Essen, Germany. The cooling capacity of this imported ice-

The 15-ton Carbonic Acid Ice Machine built for Nick Kessler by Krupp Machine Works of Essen, Germany in 1886.

making machine, driven by a 75-horse-power Hamilton Corliss engine, was equivalent to the melting of 25 tons of natural ice daily. However, in operation, it did not prove as successful as was anticipated. To be truthful, it possibly was under repair as often as it was in operation. When local mechanics gave up, Kessler sent to Essen, Germany, for an expert, Fritz Arndt.

Later, when American manufacturers perfected ice-making machines, the Krupp refrigerator was discarded and one of American make was installed.

The next few years were uneventful. The Brewery continued to grow and to brew a product favored by Montanans. Then, in 1901, Nick Kessler died. He was succeeded in the presidency by his son, Charles. Another son, Frederick was named vice president and treasurer. Charles, besides serving as president, was in charge of the brewing, a position for which he was well fitted as he was trained in the famed Wahl-Heinus Institute of Chicago.

Business grew and the old aging tanks that had served so well, were not large enough to contain the brew. In 1903, the first glass-lined storage tanks were installed. Two years later, bottled beer had assumed a larger importance, not only because of local preference but because Kessler began to distribute its beer in other towns away from Helna. To meet these conditions, in 1907 Charles Kessler installed the first Government Pipe Line between the East and the Pacific Coast.

It was along about this time, that the management of the Brewery became interested in delivering beer by automotive means rather than the big brewery

6

7

On this page, the brewery claims that it likely installed the first ice machine between Chicago and the Pacific coast, a 15-ton carbonic acid ice machine built by Krupp Machine Works in Essen, Germany. (Courtesy of Lozar's Montana Brewery Museum.)

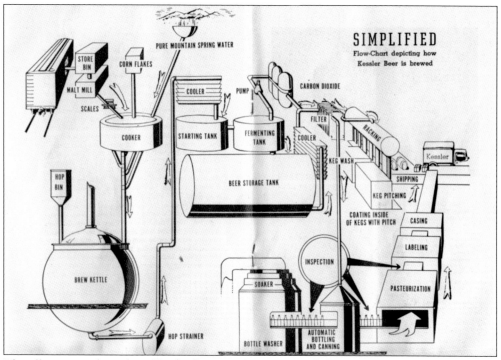

This illustration shows how Kessler Brewing Company made beer, from pure mountain spring water to loading onto the truck. (Courtesy of Lozar's Montana Brewery Museum.)

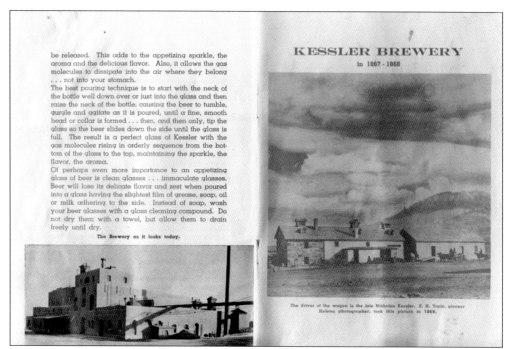

The final pages of the document compare the appearance of the Kessler brewery in 1868 and "as it looks today." (Courtesy of Lozar's Montana Brewery Museum.)

the Brewing

The first impression you receive when you step into the brewery is its spotless cleanliness. Every piece of equipment: machinery, vats, pipes, are shining. The operation efficient, controlled. The next thing you will notice will be the softly scented air . . . the sweet odor of malt liquor mingled with the tang of crushed hops and the rich fragrance of yeast. You seem to know that here men are working with wholesome, natural materials. Nowhere do you experience the odor of synthetic or imitation products.

The sweet odor is from the malt that came originally from a sun-drenched barley field. American malt is used as it enjoys a higher protein content than malt from Europe. The premium grains are harvested at just the right moment, for barley destined to be made into malt, is a delicate grain. The grain is speeded to the malting plant and, after being screened and steeped in pure water, is dried. Now, the barley malt has ripened and has evolved the properties that allow the malt starch to be converted into malt sugar. It is shipped directly to the brewery where it is stored.

Master brewers at Kessler's will measure the barley malt in the same careful way a housewife measures the ingredients for a prize cake . . . exact; precise. The malt, along with pure mountain spring water from Kessler's private springs, goes into a closed vessel called

15

Upper Left—Charles N. Kessler, Second President and still Vice-President in charge of brewing.

Upper Right — Frederick E. Kessler, Third President, now deceased.

Right— Marc W. Buterbaugh, present President of Kessler Brewing Company.

These were the three major movers for Kessler Brewing, aside from Nickolas Kessler. Pictured are Charles Kessler (top left), Frederick Kessler (top right), and Frederick's son-in-law Marc Buterbaugh, who oversaw the final days of the brewery. (Courtesy of Lozar's Montana Brewery Museum.)

Pictured here is a small bank bag used to make deposits from the brewery. (Courtesy of Lozar's Montana Brewery Museum.)

The following pages include the most complete collection of Kessler Brewing Company calendar pages featuring risqué pin-up models. The collection spans between 1952 and 1954. (Courtesy of Lozar's Montana Brewery Museum.)

43

The calendar models were drawn by famed pin-up artist Earl Moran, whose depictions of Marilyn Monroe helped launch her career. (Courtesy of Lozar's Montana Brewery Museum.)

Each month features a brief paragraph on Montana history. This one ties into the "Brewed From Mountain Spring Water" tagline with a discussion of the glacial ice that once covered much of the state. (Courtesy of Lozar's Montana Brewery Museum.)

Conscience gets a lot of credit that belongs to cold feet.

EARL MORAN

KESSLER BREWING CO.
HELENA, MONTANA

Then came the ice . . . the continental ice sheet slid down from the west of Hudson Bay and covered all of the northern part of Montana east of the Rockies. It filled the stream beds with rock debris and started the ancient Missouri cutting new channels. When it receded, it left its tracks in the form of ridges and moraines. Since then the changes in Montana's surface have been the much less violent kind that come from erosion.

But Kessler beer never changes — it's skillfully brewed to a time-tested formula — always delightfully satisfying. Serve it for repeat business. Your distributor always has it.

KESSLER

•BEER•

Brewed From Mountain Spring Water

JANUARY 1953	FEBRUARY 1953	MARCH 1953
S M T W T F S	S M T W T F S	S M T W T F S
1 2 3	1 2 3 4 5 6 7	1 2 3 4 5 6 7
4 5 6 7 8 9 10	8 9 10 11 12 13 14	8 9 10 11 12 13 14
11 12 13 14 15 16 17	15 16 17 18 19 20 21	15 16 17 18 19 20 21
18 19 20 21 22 23 24	22 23 24 25 26 27 28	22 23 24 25 26 27 28
25 26 27 28 29 30 31		29 30 31

Montana — The Treasure State

No kisses, and a girl and her honey are soon parted.

EARL MORAN

KESSLER BREWING CO.
HELENA, MONTANA

'The Land of the Shining Mountains' — that's what La Verendrye exclaimed when he first saw the snow-capped mountains to the west glistening in the sunlight like burnished silver. On New Year's Day, 1743, this adventurous Frenchman entered the southeastern corner of Montana and became the first white man to reach the area.

Brewed from cool, clear, sparkling mountain spring water, choice sun-ripened grains and tangy hops — Kessler beer is first among the fine brews. It pleases everyone who enjoys good beer. Serve it always. Order from your distributor.

KESSLER
·BEER·

Brewed From Mountain Spring Water

MARCH 1953								APRIL 1953								MAY 1953					
S	M	T	W	T	F	S		S	M	T	W	T	F	S		S	M	T	W	T	F	S
1	2	3	4	5	6	7					1	2	3	4						1	2
8	9	10	11	12	13	14		5	6	7	8	9	10	11		3	4	5	6	7	8	9
15	16	17	18	19	20	21		12	13	14	15	16	17	18		10	11	12	13	14	15	16
22	23	24	25	26	27	28		19	20	21	22	23	24	25		17	18	19	20	21	22	23
29	30	31						26	27	28	29	30				24/31	25	26	27	28	29	30

Montana — The Treasure State

Featured at the bottom of this and many other calendar pages is Montana's official nickname, the "Treasure State." Its alternate nickname, "Big Sky Country," did not become popular until a state highway department program in the 1960s. (Courtesy of Lozar's Montana Brewery Museum.)

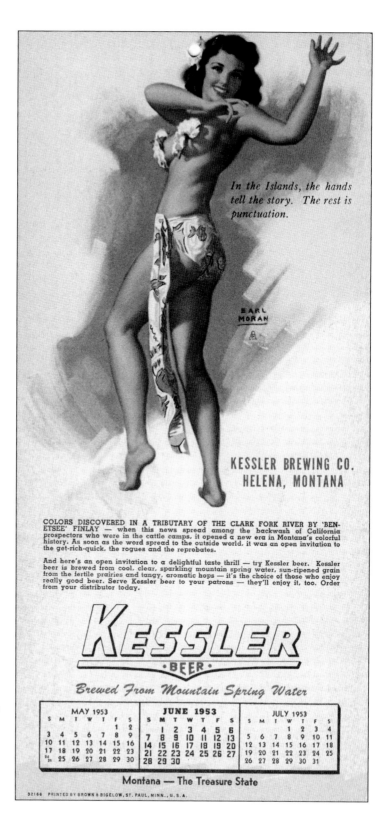

In the Islands, the hands tell the story. The rest is punctuation.

EARL MORAN

KESSLER BREWING CO.
HELENA, MONTANA

COLORS DISCOVERED IN A TRIBUTARY OF THE CLARK FORK RIVER BY 'BEN-ETSEE' FINLAY — when this news spread among the backwash of California prospectors who were in the cattle camps, it opened a new era in Montana's colorful history. As soon as the word spread to the outside world, it was an open invitation to the get-rich-quick, the rogues and the reprobates.

And here's an open invitation to a delightful taste thrill — try Kessler beer. Kessler beer is brewed from cool, clear, sparkling mountain spring water, sun-ripened grain from the fertile prairies and tangy, aromatic hops — it's the choice of those who enjoy really good beer. Serve Kessler beer to your patrons — they'll enjoy it, too. Order from your distributor today.

KESSLER
· BEER ·

Brewed From Mountain Spring Water

MAY 1953							JUNE 1953							JULY 1953					
S	M	T	W	T	F	S	S	M	T	W	T	F	S	S	M	T	W	T	F	S
					1	2		1	2	3	4	5	6				1	2	3	4
3	4	5	6	7	8	9	7	8	9	10	11	12	13	5	6	7	8	9	10	11
10	11	12	13	14	15	16	14	15	16	17	18	19	20	12	13	14	15	16	17	18
17	18	19	20	21	22	23	21	22	23	24	25	26	27	19	20	21	22	23	24	25
24 31	25	26	27	28	29	30	28	29	30					26	27	28	29	30	31	

Montana — The Treasure State

32166 PRINTED BY BROWN & BIGELOW, ST. PAUL, MINN., U.S.A.

These calendars were intended primarily for wholesale customers of the brewery, as indicated by the line "Serve Kessler beer to your patrons—they'll enjoy it." Bar owners no doubt found that many of their customers also enjoyed the calendars themselves. (Courtesy of Lozar's Montana Brewery Museum.)

The only Chaps a girl can trust are those she wears.

EARL MORAN USA

KESSLER BREWING CO.
HELENA, MONTANA

Montana's first pay discovery of placer gold was made at Grasshopper Creek by John White in 1862. Here the camp of Bannack came into being. The influx of fortune seekers was so great that when Montana became a territory in May, 1864, Bannack was made the first capital.

Another 'first' you'll remember is the first time you try Kessler beer. A cool glass of this sparkling, pale amber brew is a taste thrill you'll not soon forget. Thrill your customers with Kessler beer. Order from your distributor today.

KESSLER
·BEER·

Brewed From Mountain Spring Water

JUNE 1953						
S	M	T	W	T	F	S
	1	2	3	4	5	6
7	8	9	10	11	12	13
14	15	16	17	18	19	20
21	22	23	24	25	26	27
28	29	30				

JULY 1953						
S	M	T	W	T	F	S
			1	2	3	4
5	6	7	8	9	10	11
12	13	14	15	16	17	18
19	20	21	22	23	24	25
26	27	28	29	30	31	

AUGUST 1953						
S	M	T	W	T	F	S
						1
2	3	4	5	6	7	8
9	10	11	12	13	14	15
16	17	18	19	20	21	22
23 30	24 31	25	26	27	28	29

Montana — The Treasure State

Each month features a slightly different description of Kessler beer. In July 1953, customers were promised, "a cool glass of this sparkling, pale amber brew is a taste thrill you'll not soon forget." (Courtesy of Lozar's Montana Brewery Museum.)

Customers were promised that Kessler was producing a steady supply of beer, perhaps to reassure bar owners worried about running low: "Serve Kessler beer today and every day. Your distributor can supply it." (Courtesy of Lozar's Montana Brewery Museum.)

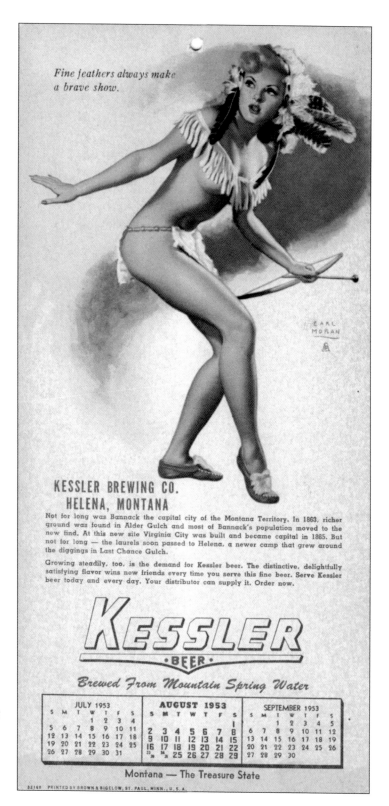

Fine feathers always make a brave show.

EARL MORAN

KESSLER BREWING CO.
HELENA, MONTANA

Not for long was Bannack the capital city of the Montana Territory. In 1863, richer ground was found in Alder Gulch and most of Bannack's population moved to the new find. At this new site Virginia City was built and became capital in 1865. But not for long — the laurels soon passed to Helena, a newer camp that grew around the diggings in Last Chance Gulch.

Growing steadily, too, is the demand for Kessler beer. The distinctive, delightfully satisfying flavor wins new friends every time you serve this fine beer. Serve Kessler beer today and every day. Your distributor can supply it. Order now.

KESSLER
• BEER •

Brewed From Mountain Spring Water

	JULY 1953							AUGUST 1953							SEPTEMBER 1953				
S	M	T	W	T	F	S	S	M	T	W	T	F	S	S	M	T	W	T	F	S
			1	2	3	4							1			1	2	3	4	5
5	6	7	8	9	10	11	2	3	4	5	6	7	8	6	7	8	9	10	11	12
12	13	14	15	16	17	18	9	10	11	12	13	14	15	13	14	15	16	17	18	19
19	20	21	22	23	24	25	16	17	18	19	20	21	22	20	21	22	23	24	25	26
26	27	28	29	30	31		23/30	24/31	25	26	27	28	29	27	28	29	30			

Montana — The Treasure State

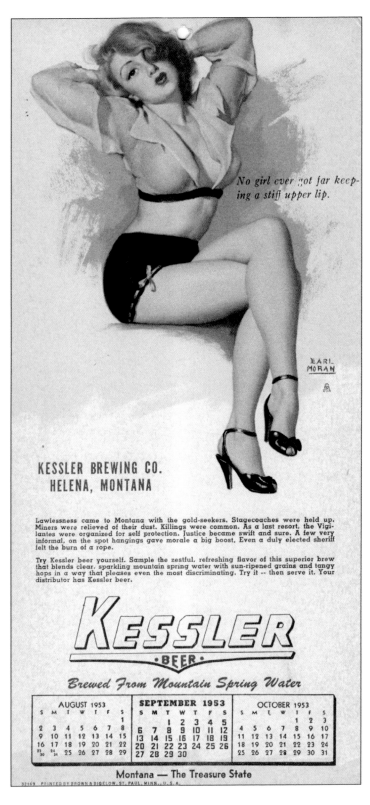

No girl ever got far keeping a stiff upper lip.

EARL MORAN

KESSLER BREWING CO.
HELENA, MONTANA

Lawlessness came to Montana with the gold-seekers. Stagecoaches were held up. Miners were relieved of their dust. Killings were common. As a last resort, the Vigilantes were organized for self protection. Justice became swift and sure. A few very informal, on the spot hangings gave morale a big boost. Even a duly elected sheriff felt the burn of a rope.

Try Kessler beer yourself. Sample the zestful, refreshing flavor of this superior brew that blends clear, sparkling mountain spring water with sun-ripened grains and tangy hops in a way that pleases even the most discriminating. Try it -- then serve it. Your distributor has Kessler beer.

KESSLER
·BEER·

Brewed From Mountain Spring Water

AUGUST 1953	SEPTEMBER 1953	OCTOBER 1953
S M T W T F S	S M T W T F S	S M T W T F S
1	1 2 3 4 5	1 2 3
2 3 4 5 6 7 8	6 7 8 9 10 11 12	4 5 6 7 8 9 10
9 10 11 12 13 14 15	13 14 15 16 17 18 19	11 12 13 14 15 16 17
16 17 18 19 20 21 22	20 21 22 23 24 25 26	18 19 20 21 22 23 24
23 30 24 31 25 26 27 28 29	27 28 29 30	25 26 27 28 29 30 31

Montana — The Treasure State

32169   PRINTED BY BROWN & BIGELOW, ST. PAUL, MINN., U.S.A.

The language of beer advertising may have changed since 1953, but the message remains much the same: "Sample the zestful, refreshing flavor of this superior brew . . . with sun-ripened grains and tangy hops . . . that pleases even the most discriminating." (Courtesy of Lozar's Montana Brewery Museum.)

In the last month
that "Montana —
The Treasure State"
would appear on a
calendar page, Kessler
doubled down on the
theme, calling its
product "the treasure
of beers." (Courtesy
of Lozar's Montana
Brewery Museum.)

*There is no loser in a
pillow fight.*

EARL
MORAN

KESSLER BREWING CO.
HELENA, MONTANA

Free land . . . brought the homesteaders to Montana, starting about 1910. Fences and
plows quickly transformed the land from vast grassy plains to squares of golden
grain. Truly the Treasure State -- mines rich in gold, silver, lead, zinc -- sleek fat
cattle on the grazing lands -- endless miles of grain fields -- fish-filled lakes, primitive
hunting areas -- thrilling vacation spots -- Montana has everything.

The treasure of beers -- Kessler . . . brewed from clear, sparkling mountain spring
water, tangy hops and sun-ripened grain . . . such aroma, such flavor, so satisfying.
Your customers will appreciate Kessler beer. Serve it today.

KESSLER
• BEER •

Brewed From Mountain Spring Water

OCTOBER 1953						
S	M	T	W	T	F	S
				1	2	3
4	5	6	7	8	9	10
11	12	13	14	15	16	17
18	19	20	21	22	23	24
25	26	27	28	29	30	31

NOVEMBER 1953						
S	M	T	W	T	F	S
1	2	3	4	5	6	7
8	9	10	11	12	13	14
15	16	17	18	19	20	21
22	23	24	25	26	27	28
29	30					

DECEMBER 1953						
S	M	T	W	T	F	S
		1	2	3	4	5
6	7	8	9	10	11	12
13	14	15	16	17	18	19
20	21	22	23	24	25	26
27	28	29	30	31		

Montana — The Treasure State

Kitchen and apron,
Dishpan and kettle,
They're not the bait
To get me to settle.

**Kessler Brewing Company
Helena, Montana**

Nearly ninety years ago, when Nick Kessler first opened Kessler Brewery, Lewis & Clark County was named Edgerton. (Lem Lomkin) Then the townsfolk changed it and everyone has been satisfied.

You make the change to Kessler Beer and you will be satisfied. Kessler Beer is brewed FOR Montanans . . . they like its light, dry yet robust flavor. (Jack McCarthy)

KESSLER
BREWED FROM MOUNTAIN SPRING WATER
Beer

NOVEMBER 1953							DECEMBER 1953							JANUARY 1954						
S	M	T	W	T	F	S	S	M	T	W	T	F	S	S	M	T	W	T	F	S
1	2	3	4	5	6	7			1	2	3	4	5						1	2
8	9	10	11	12	13	14	6	7	8	9	10	11	12	3	4	5	6	7	8	9
15	16	17	18	19	20	21	13	14	15	16	17	18	19	10	11	12	13	14	15	16
22	23	24	25	26	27	28	20	21	22	23	24	25	26	17	18	19	20	21	22	23
29	30						27	28	29	30	31			24/31	25	26	27	28	29	30

Look for Your Name

If it appears anywhere in this ad you'll receive a FREE case of Kessler beer if you return this card to us. Look for your name each month.

33172 PRINTED BY BROWN & BIGELOW, ST. PAUL, MINN., U.S.A.

The style of the calendar changed in December 1953 from the previous month, with a different design and Montana's "Treasure State" nickname no longer featured. (Courtesy of Lozar's Montana Brewery Museum.)

A girl like me,
Of pleasant dimension,
Tonight could go for
Some masculine attention.

Kessler Brewing Co.
Helena, Montana

SCOOP

The ROCKY MOUNTAIN HUSBANDMAN, in November of 1884, printed this terse statement: "Two men, Jones and Valentine, went hunting in the Highwood neighborhood. (Martin Toohey) Jones returned with some game, but Valentine has not, and it is suspected that Jones has murdered him." (John Rice)

No suspicions about Kessler beer, though. In the Highwood neighborhood, or all over Montana, for that matter, more people are turning to this fine, sturdy brew. Kessler beer—the beer with a flavor of its own.

KESSLER
BREWED FROM MOUNTAIN SPRING WATER
Beer

APRIL 1954							MAY 1954							JUNE 1954					
S	M	T	W	T	F	S	S	M	T	W	T	F	S	S	M	T	W	T	F	S
				1	2	3							1			1	2	3	4	5
4	5	6	7	8	9	10	2	3	4	5	6	7	8	6	7	8	9	10	11	12
11	12	13	14	15	16	17	9	10	11	12	13	14	15	13	14	15	16	17	18	19
18	19	20	21	22	23	24	16	17	18	19	20	21	22	20	21	22	23	24	25	26
25	26	27	28	29	30		23/30	24/31	25	26	27	28	29	27	28	29	30			

Look for Your Name

If it appears anywhere in this ad you'll receive a FREE case of Kessler beer if you return this card to us. Look for your name each month.

33165 PRINTED BY BROWN & BIGELOW, ST. PAUL, MINN., U. S. A.

The style change also brought a new gimmick. Each month, the names of two customers were inserted within the text. Lucky patrons whose names were featured were entitled to a free case of beer. (Courtesy of Lozar's Montana Brewery Museum.)

The peas are up,
The carrots are fine.
I'm eager to be
Your clinging vine.

Last Chance Gulch -- scene of the final desperate hopes of the Georgians, unlucky prospectors. They struck it rich. The Gulch was one of the richest gold bearing deposits in Montana. Hundreds of miners swarmed in, and Helena was founded in the fall of 1864. (W. A. Disbrow) The main street is laid out along the original gulch. (Morris Davis)

You'll strike it rich with Kessler beer. People who know their beer buy the best. And in Montana, it's KESSLER beer, favored since 1865.

Lewis and Clark Brewing Company, which is open today in Helena, launched a line of retro-themed advertisements inspired by these calendar pages and featuring similar pin-up art. (Courtesy of Lozar's Montana Brewery Museum.)

In August 1954, in their 89th year of business, Kessler Brewing Company clams here that "absolutely no dissatisfaction from any customer has been reported." (Courtesy of Lozar's Montana Brewery Museum.)

Kessler Brewing Co.
Helena, Montana

See what I'm wearing?
A cute little vest.
Hope it'll bring
A few men west.

Social Note: A neck-tie party, arranged by the Vigilantes for a certain Montana road agent chief, took place as planned. ('Brother' Spillum) The hanging was postponed until after a turkey dinner given by the chief for some of the Vigilantes. A good time was had by all. (James Mulry)

If the chief had had the opportunity to serve Kessler beer along with that turkey, chances are he'd still be alive to drink it today. Kessler beer turns any social gathering into a real occasion. Order some soon.

KESSLER
BREWED FROM MOUNTAIN SPRING WATER
Beer

AUGUST 1954						
S	M	T	W	T	F	S
1	2	3	4	5	6	7
8	9	10	11	12	13	14
15	16	17	18	19	20	21
22	23	24	25	26	27	28
29	30	31				

SEPTEMBER 1954						
S	M	T	W	T	F	S
			1	2	3	4
5	6	7	8	9	10	11
12	13	14	15	16	17	18
19	20	21	22	23	24	25
26	27	28	29	30		

OCTOBER 1954						
S	M	T	W	T	F	S
					1	2
3	4	5	6	7	8	9
10	11	12	13	14	15	16
17	18	19	20	21	22	23
24 31	25	26	27	28	29	30

Look for Your Name
If it appears anywhere in this ad you'll receive a FREE case of Kessler beer if you return this card to us. Look for your name each month.
33159 PRINTED BY BROWN & BIGELOW, ST. PAUL, MINN., U. S. A.

These calendars appeared about four years before the brewery folded. This form of advertising, while popular among certain customers, could not save the brewery. (Courtesy of Lozar's Montana Brewery Museum.)

Nice little sweater,
So warm and cozy.
Won't hide much
On our little Rosie.

Kessler Brewing Co.
Helena, Montana

The Curry Brothers, Billy the Kid, Hank Plummer -- all bad actin' hombres in their day, and all snakes in the grass. But the two werst ones known to have existed in Montana were Triceratops and Tyrannosaurus Rex. (Barney Baker) Fossils of these dinosaurs indicate that Montana's geologic past was as garishly colorful as its history book past. (Adolph Simek)

Kessler beer started back in Montana's history book past. Since 1865, this golden beverage has been pleasing Montana's palate. Order some from your dealer today.

KESSLER
BREWED FROM MOUNTAIN SPRING WATER
Beer

SEPTEMBER 1954							OCTOBER 1954							NOVEMBER 1954						
S	M	T	W	T	F	S	S	M	T	W	T	F	S	S	M	T	W	T	F	S
			1	2	3	4						1	2		1	2	3	4	5	6
5	6	7	8	9	10	11	3	4	5	6	7	8	9	7	8	9	10	11	12	13
12	13	14	15	16	17	18	10	11	12	13	14	15	16	14	15	16	17	18	19	20
19	20	21	22	23	24	25	17	18	19	20	21	22	23	21	22	23	24	25	26	27
26	27	28	29	30			24 31	25	26	27	28	29	30	28	29	30				

Look for Your Name
If it appears anywhere in this ad you'll receive a FREE case of Kessler beer if you return this card to us. Look for your name each month.
33170 PRINTED BY BROWN & BIGELOW, ST. PAUL, MINN., U.S.A.

This month covered a different kind of treasure that Montana is also known for—dinosaur fossils. Likely few beer ads today compare Billy the Kid to Tyrannosaurus Rex. (Courtesy of Lozar's Montana Brewery Museum.)

Send a signal,
Beat a drum.
Little Hummingbird
Has a man on the run!

Kessler Brewing Co.
Helena, Montana

Boomtown of the gold rush days, Virginia City, was originally named Varina in honor of Mrs. Jefferson Davis, wife of the president of the Confederacy. (Eddy Coyle) The Johnnie Rebs and the damyankees got to shooting it out in the main street of town, so the name had to be changed.

The best reason there is for drinking Kessler beer is very simple. It tastes good. (Rex Thomas) It's brewed by people who know fine beer, for customers who know fine beer. Your distributor has it. Find out for yourself why more and more people are drinking Kessler beer.

KESSLER

BREWED FROM MOUNTAIN SPRING WATER

Beer

OCTOBER 1954						
S	M	T	W	T	F	S
					1	2
3	4	5	6	7	8	9
10	11	12	13	14	15	16
17	18	19	20	21	22	23
24/31	25	26	27	28	29	30

NOVEMBER 1954						
S	M	T	W	T	F	S
	1	2	3	4	5	6
7	8	9	10	11	12	13
14	15	16	17	18	19	20
21	22	23	24	25	26	27
28	29	30				

DECEMBER 1954						
S	M	T	W	T	F	S
			1	2	3	4
5	6	7	8	9	10	11
12	13	14	15	16	17	18
19	20	21	22	23	24	25
26	27	28	29	30	31	

Look for Your Name

If it appears anywhere in this ad you'll receive a FREE case of Kessler beer if you return this card to us. Look for your name each month.

33171 PRINTED BY BROWN & BIGELOW, ST. PAUL, MINN. U.S.A.

The last calendar page features perhaps the simplest and most direct reason to drink Kessler beer: "It tastes good." (Courtesy of Lozar's Montana Brewery Museum.)

Perhaps a slightly controversial name for a beer, Lorelei was very popular. The name comes from a rocky headland on the Rhine River. Legends say that a maiden named Lorelei lives on the rock and lures fishermen to their death with her song. (Courtesy of Helenahistory.org.)

Pictured here is the Kessler Brewery, as engraved for an 1890 perspective map of Helena. (Courtesy of Helenahistory.org.)

In October and November 1935, Helena suffered a series of strong earthquakes, accompanied by hundreds of small aftershocks. Damage was severe and widespread. The Kessler smokestack was cracked and the boiler room damaged in one of the early quakes. A Northern Pacific locomotive was used to power the brewery for a time, as shown above. (Both, courtesy of Helenahistory.org.)

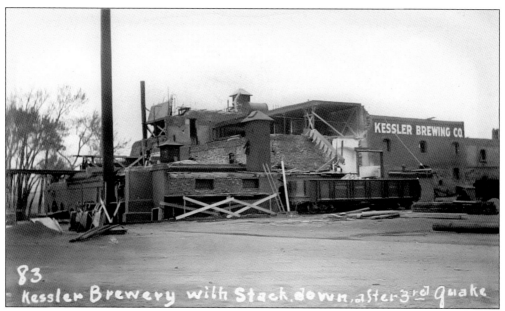

After the third earthquake, the smokestack was completely ruined. (Courtesy of Helenahistory.org.)

Pictured here are Kessler employees on July 21, 1949. Their facial hair was grown for a citywide Golden Canyon Days celebration. (Courtesy of Helenahistory.org.)

FIRST CHRISTMAS IN HAWAII.

Kessler Brewing Company was not the only brewery in Helena to achieve notoriety. Capital Brewing Company was also started in 1865. It later became a popular theater site. In this early advertisement, the suggestion was to resist temptation while away on military duty. (Courtesy of Lozar's Montana Brewery Museum.)

The Olde Brewery Theatre operated in the site of the former Capital Brewery building from 1954 to 1972. The building was destroyed in July 1973 as part of the urban-renewal program. A few years after the theater opened, in September 1957, Helena native and actor Gary Cooper visited the theater and Helena. It was his last visit to Helena before he died in 1961. (Courtesy of Helenahistory.org.)

Though not a brewery, any mention of Helena would be remiss if it did not include the Central Beer Hall on 118 West Main Street. Before it was demolished in 1971 as part of the urban-renewal program, it was Helena's oldest bar. According to old fire-insurance maps, the building was constructed as a saloon between 1884 and 1888. The Central Beer Hall was started by German native Henry Rossmann (1865–1942). During Prohibition, the business was listed in the city directory as "Henry Rossman, Soft Drinks." (Courtesy of Helenahistory.org.)

Four

GREAT FALLS

Great Falls was on the map for Lewis and Clark when they explored in the early 19th century. It had everything—the headwaters of the Missouri, ample wildlife, and rivers to trade furs on. As the 20th century approached, it was clear that beer would be woven into the tapestry of the city's history. Great Falls had some of the earliest breweries in the state, and it had the lone survivor, Great Falls Breweries Inc., which lasted until 1968. When it comes to Montana's brewing traditions, Great Falls fills a substantial chapter.

The first brewery in Great Falls was Volk Brewery, which opened in 1888 on Upper River Road. It was founded by Christian Volk Sr. and Nicholas Volk. (Courtesy of Lozar's Montana Brewery Museum.)

Volk Brewery specialized in colorful, dynamic label art for its bottles. (Both, courtesy of Lozar's Montana Brewery Museum.)

The brewery was built from native stone and wood, and its beers were delivered throughout the Great Falls area and Sun River Valley. (Courtesy of Lozar's Montana Brewery Museum.)

One of the most unique pieces of Volk Brewery breweriana to survive is this small leather-wrapped broom, which was carried by umpires to clear home plate during baseball games. (Courtesy of Lozar's Montana Brewery Museum.)

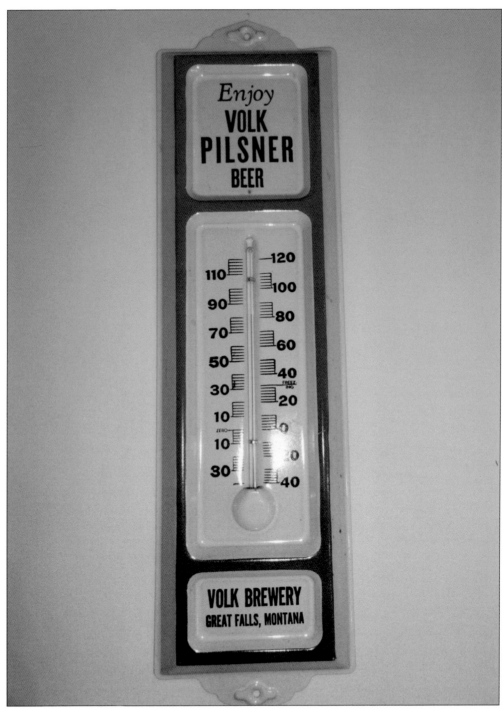

Unfortunately, the Volk Brewery burned down in 1894, leaving only its stone foundation. The year that the Volk Brewery burned, the Montana Brewing Company released advertisements that read, "We guarantee our Beers to be Strictly Pure, Brewed from the Choicest Brands of Imported Hops and Montana Malt . . . All orders for keg or bottle beer promptly attended to and delivered to any part of the city." (Courtesy of Lozar's Montana Brewery Museum.)

Before American Brewing Company came to life right before Prohibition, it produced and sold beer as the Montana Brewing Company. The brewery was a massive complex on the west end of First Avenue North, on the bank of the Missouri River in Great Falls. As the American Brewing Company, the brewery released colorful advertisements. (Courtesy of Lozar's Montana Brewery Museum.)

This unique label featured both American Brewing Company and Montana Brewing Company, but it was produced under Great Falls Breweries Inc. This label was short-lived because it looked too German at a time when the world was at war with Germany. The beer became known as Great Falls Beer: Bohemian Style Lager. (Courtesy of Lozar's Montana Brewery Museum.)

Another American Beer advertisement promised men that a tête-à-tête with a woman would go much better if they were both drinking the American brand of beer, because it "furnishes animation, sharpens the wit and makes conversation flow." Once Prohibition went into effect, however, both the American brand and colorful advertisements were gone. In 1933, when brewing was legal again, Emil Sick stepped in and bought interest in the former brewery and launched Great Falls Breweries Inc. (Courtesy of Lozar's Montana Brewery Museum.)

Beer baron Emil Sick was ready to make beer when Prohibition was repealed. His signature beer was a Bohemian-style lager, but he quickly changed it to Great Falls Select because of pressures in Europe. Hitler was coming into power, and breweries in the United States were asked to divest themselves of anything with a German tone. (Courtesy of Lozar's Montana Brewery Museum.)

Emil Sick managed the brewery until he sold it to locals in Great Falls in 1949. The new owners kept brewing Great Falls Select because, even though it suffered a poor reputation at times, it was still one of the most famous Montana beers of the 20th century. (Courtesy of Lozar's Montana Brewery Museum.)

The owners Emil Sick sold the brewery to introduced three new beers, including Big Sky, Frontier Town, and Western. In 1966, those owners sold their interest to the Blitz-Weinhard Brewing Company, based in Portland, Oregon. Blitz-Weinhard brewed Great Falls Select in Great Falls for two more years until it shut down the brewery and moved production to its brewery in Oregon. The brand was produced until around 1980, when Blitz-Weinhard cited a lack of funds to properly promote the beer and retired it. (Courtesy of Lozar's Montana Brewery Museum.)

During its long tenure, there was no shortage of souvenirs, household items, and advertising gimmicks that bore the Great Falls Breweries and Great Falls Select name. Pictured here is a pink toothpick holder. (Courtesy of Lozar's Montana Brewery Museum.)

Anyone who could make beer could make a living in Great Falls, even at the turn of the 20th century. According to University of Great Falls professor William Furdell, a brewer made $3.50 a day, a smelter worker made between $3 and $4.50, and a cook made $1.25 plus meals in 1900. (Courtesy of Lozar's Montana Brewery Museum.)

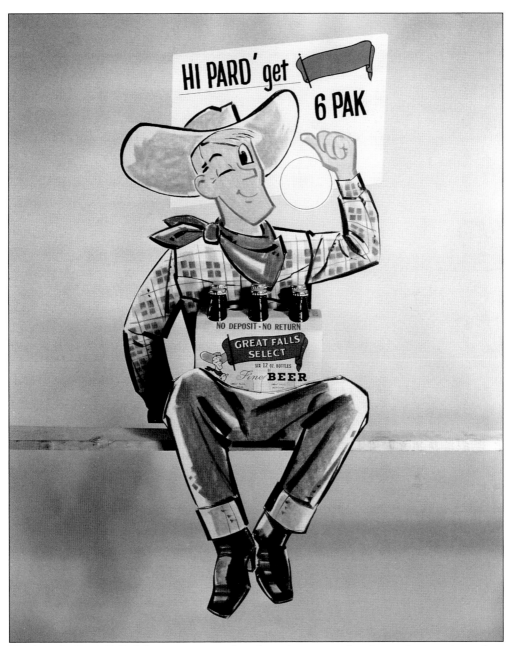

The love for Great Falls Select was universal, and its advertising often featured an iconic cowboy. What better image to sell Montana beer than this? (Courtesy of Lozar's Montana Brewery Museum.)

The combination of beer and food is nothing new. Great Falls Breweries capitalized on the trend of pairing beer with food and cooking with beer for years. The brewery often stocked cards on its bottles with food suggestions, as seen here. (Courtesy of Lozar's Montana Brewery Museum.)

GREAT FALLS SELECT
BARBEQUE SAUCE

½ cup Great Falls Select

½ cup catsup

1 teaspoon Worcestershire sauce

1 teaspoon sugar

2 dashes of Tabasco

½ teaspoon monosodium glutumate

Boil gently for 5 minutes. slice frankfurters and cook for 3 minutes. Meatballs may be cooked in this sauce, or it can be served over hamburgers.

special <u>lively</u> flavor-
THAT CATERS TO THE TASTE OF THE WEST

GREAT FALLS BREWERIES, INC., GREAT FALLS, MONTANA

In addition to reminders on the bottles, Great Falls Breweries published a series of recipes, including this one for Great Falls Select barbecue sauce with "special lively flavor that caters to the taste of the west." (Courtesy of Lozar's Montana Brewery Museum.)

"GREAT" GLAZED HAM

Buy the type and size ham you like. Cook it, if necessary, and drain it well. Place the ham in a roasting pan and bake in a 350° oven 45 minutes for a half ham, 1 hour for a whole ham. Baste with 1 cup GREAT FALLS SELECT. Remove the ham from the oven and score the fat diagonally in a crisscross pattern. Brush the ham with either of the following:

1¼ cups brown sugar
3 tablespoons bread crumbs
1 teaspoon dry mustard
1 cup GREAT FALLS SELECT

Mix the sugar, bread crumbs, and mustard; add enough beer to make a paste.
1 cup applesauce
⅔ cup brown sugar
¼ teaspoon nutmeg
½ teaspoon cinnamon
½ cup GREAT FALLS SELECT

Mix all the ingredients together and spread over the ham. In either case, stud the ham with cloves and bake 30 minutes longer. Baste occasionally if the bread-crumb mixture is used. (NOTE: a canned ham may be prepared in the same way, but first cut it in ½-inch slices. Re-assemble and tie it with white string. Spread either of the glazes over it and proceed as directed.)

"Great" glazed ham was a seasonal suggestion that called for basting the ham with one cup of Great Falls Select. (Courtesy of Lozar's Montana Brewery Museum.)

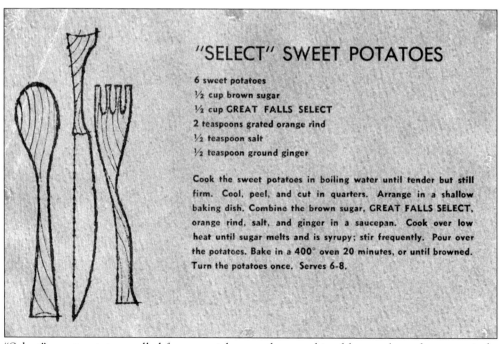

"SELECT" SWEET POTATOES

6 sweet potatoes
½ cup brown sugar
½ cup GREAT FALLS SELECT
2 teaspoons grated orange rind
½ teaspoon salt
½ teaspoon ground ginger

Cook the sweet potatoes in boiling water until tender but still firm. Cool, peel, and cut in quarters. Arrange in a shallow baking dish. Combine the brown sugar, GREAT FALLS SELECT, orange rind, salt, and ginger in a saucepan. Cook over low heat until sugar melts and is syrupy; stir frequently. Pour over the potatoes. Bake in a 400° oven 20 minutes, or until browned. Turn the potatoes once. Serves 6-8.

"Select" sweet potatoes called for a special sauce that combined beer, salt, and orange rinds. (Courtesy of Lozar's Montana Brewery Museum.)

BAKING POWDER IN BEER BRIOCHE

2 cups sifted flour
½ teaspoon salt
⅓ cup sugar
1 tablespoon baking powder
4 eggs
⅓ cup GREAT FALLS SELECT
¾ cup butter, softened

Sift together the flour, salt, sugar, and baking powder. Beat the eggs until light and fluffy. Alternately add the flour mixture and the GREAT FALLS SELECT. Beat in the butter. Cover and let stand in a cool place (not the refrigerator) overnight. (The batter may be baked after 1 hour, but the longer time improves the flavor.) Preheat the oven to 375°. Bake in 24 buttered muffin tins 15 minutes or in a 10-inch loaf pan 35 minutes, or until browned. The beer supplies the flavor and lightness of yeast, without the trouble of kneading and rising.

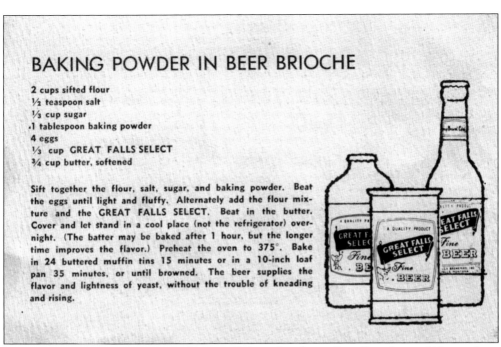

Perhaps not the most appetizing of names, Baking Powder in Beer Brioche could be made as muffins or in a loaf pan. (Courtesy of Lozar's Montana Brewery Museum.)

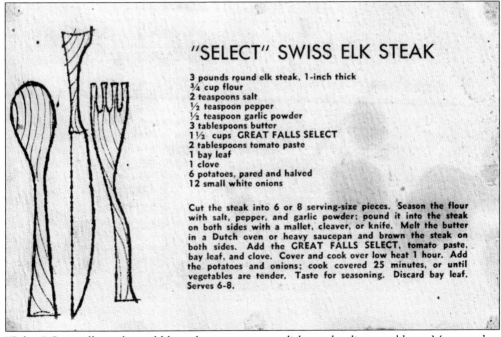

"SELECT" SWISS ELK STEAK

3 pounds round elk steak, 1-inch thick
¾ cup flour
2 teaspoons salt
½ teaspoon pepper
½ teaspoon garlic powder
3 tablespoons butter
1½ cups GREAT FALLS SELECT
2 tablespoons tomato paste
1 bay leaf
1 clove
6 potatoes, pared and halved
12 small white onions

Cut the steak into 6 or 8 serving-size pieces. Season the flour with salt, pepper, and garlic powder; pound it into the steak on both sides with a mallet, cleaver, or knife. Melt the butter in a Dutch oven or heavy saucepan and brown the steak on both sides. Add the GREAT FALLS SELECT, tomato paste, bay leaf, and clove. Cover and cook over low heat 1 hour. Add the potatoes and onions; cook covered 25 minutes, or until vegetables are tender. Taste for seasoning. Discard bay leaf. Serves 6-8.

"Select" Swiss elk steak would have been a common dish on the dinner table, as Montana has its own long history tied to hunting elk, deer, and other game. This was a real meat-and-potatoes dish. (Courtesy of Lozar's Montana Brewery Museum.)

"GREAT" FLAPJACKS

1½ cups sifted flour
¼ teaspoon salt
1½ teaspoons double-action baking powder
2 teaspoons sugar
2 egg yolks
3 tablespoons melted butter
½ cup GREAT FALLS SELECT
⅔ cup milk
1 egg white, stiffly beaten

Sift together the flour, salt, baking powder, and sugar. Beat together the egg yolks, butter, GREAT FALLS SELECT, and milk. Add to the flour mixture, stirring until smooth. Fold in the egg white. Drop by the tablespoonful onto a hot greased griddle or skillet. Bake until bubbles cover the top, then turn and bake until browned on other side. Don't turn more than once. Serve with hot syrup or honey. Makes about 18 3-inch pancakes.

Beer was not just for lunch and dinner. Here it makes an appearance for breaks. "Great" flapjacks only used four ounces of beer in the recipe, and one can wonder what happened to the rest of an opened can. (Courtesy of Lozar's Montana Brewery Museum.)

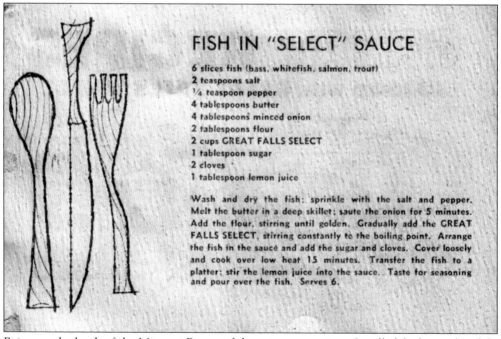

FISH IN "SELECT" SAUCE

6 slices fish (bass, whitefish, salmon, trout)
2 teaspoons salt
¼ teaspoon pepper
4 tablespoons butter
4 tablespoons minced onion
2 tablespoons flour
2 cups GREAT FALLS SELECT
1 tablespoon sugar
2 cloves
1 tablespoon lemon juice

Wash and dry the fish; sprinkle with the salt and pepper. Melt the butter in a deep skillet; saute the onion for 5 minutes. Add the flour, stirring until golden. Gradually add the GREAT FALLS SELECT, stirring constantly to the boiling point. Arrange the fish in the sauce and add the sugar and cloves. Cover loosely and cook over low heat 15 minutes. Transfer the fish to a platter; stir the lemon juice into the sauce. Taste for seasoning and pour over the fish. Serves 6.

Being on the bank of the Missouri River, a fish recipe was a given. It called for bass, whitefish, trout, or salmon, as well as two cups of Great Falls Select, assuming any beer was left over after a day of catching the fish. (Courtesy of Lozar's Montana Brewery Museum.)

Like the recipes for elk and trout, Great Falls Breweries aligned itself with outdoor recreationists and sportsmen of Montana. They published a guide full of outdoor tips, complete with a cartoon picturing a bear catching an arctic grayling. (Courtesy of Lozar's Montana Brewery Museum.)

Great Falls Select was available on tap, in cans, and in bottles, as illustrated here. Again, the motif of beer in the outdoors was common through Great Falls Breweries' run. (Courtesy of Lozar's Montana Brewery Museum.)

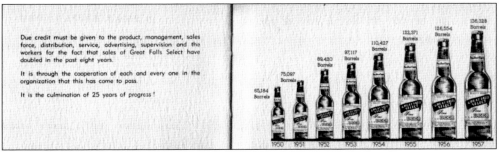

This interesting publication put out by Great Falls Breweries shows how production more than doubled between 1950 and 1957, yet the brewery would close its doors 11 years later in 1968. (Courtesy of Lozar's Montana Brewery Museum.)

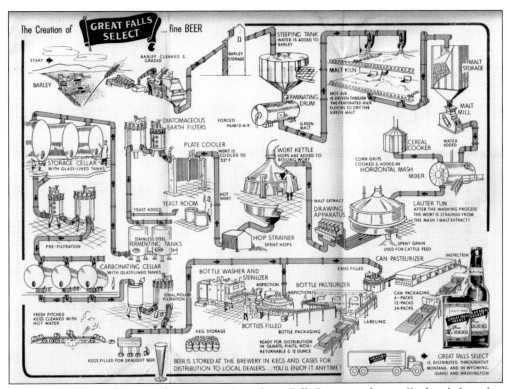

In the 1950s, science kept people's interest. Here, Great Falls Breweries shows off a detailed graphic that outlines all the technology behind making each bottle of Great Falls Select. (Courtesy of Lozar's Montana Brewery Museum.)

SINCE the relegalization of beer in 1933, Great Falls Breweries, Inc. has consistently endeavored to be a good citizen of this community, this state and nation. In so doing, **Great Falls Breweries** has kept pace with the forward advances of the brewing industry and has established and maintained one of the most modern plants in the Northwest. For instance: stainless steel is the very latest in piping material for use in brewing. During this tour you will see the stainless steel transfer lines which Great Falls Breweries has introduced in its continuing program of plant modernization. Already it has many stainless steel fermenting tanks. You are invited to inspect these and many other innovations during your tour of the plant.

The annual output is about 4,340,000 gallons of Great Falls Select and is distributed and sold in Montana, Idaho, Wyoming and Washington. Other statistics which will interest you include: a staff of more than 125 employees, a payroll of over $577,000 in wages, with a profit-sharing plan for employees; each year Great Falls Breweries buys more than $537,000 in local services and supplies, purchases over 4,000,000 pounds of Montana barley, and pays more than $1,669,000 in state and federal taxes. Careful brewing and years of experience have improved the quality of every glass and can and bottle of Great Falls Select. We hope you will see this during your tour and will agree with the many Montanans who enjoy Great Falls Select and have made it . . .

"First in sales in Montana"

This handbill touts Great Falls Select as "First in Sales in Montana," telling of production equal to 4.3 million gallons of beer sold in Montana, Washington, Idaho, and Wyoming. (Courtesy of Lozar's Montana Brewery Museum.)

Five

THE MANY OTHERS

It would take volumes of books to illustrate even half of Montana's rich brewing history, a history that spans more than 150 years. One visit to Lozar's Montana Brewery Museum will prove this to be true. Thanks to the dedicated work of Steve Lozar, we know today that Montana welcomed at least 197 distinct breweries over the years. There are over 90 breweries open today, with more on the way. No one knows how many the state *can* hold, but it holds them well.

What follows is a selection of breweries that vary in location but were quite important for their roles in Montana brewery history. From Kalispell to Billings, big breweries, small breweries, and brewery owners with ties to national brands, all had a part to play in writing the script.

Brewery-Virginia City Mont.

The H.S. Gilbert in Virginia City, Montana, is widely claimed to be the state's first brewery, although Steve Lozar's research can point to a few others that came before. Regardless, the brewery is iconic, and a portion of it still stands and is still photographed today. It is the home of the production Brewery Follies, and beer is available inside the brewery. This photograph is not the common shot of the building's exterior. Instead, it shows a surly character who is either waiting for a poker game or just wants to be left alone. (Courtesy of Lozar's Montana Brewery Museum.)

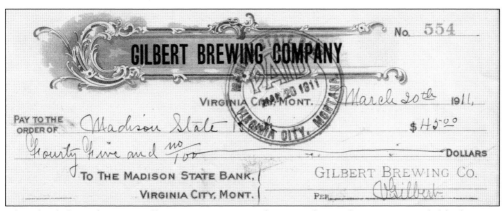

This brewery license, which cost $20 in 1912, was issued when Henry S. Gilbert's daughter Valentine Gilbert was in charge of the brewery. H.S. Gilbert died in 1902. Valentine operated the brewery until Prohibition forced it to close, and it could not rebound. (Courtesy of Lozar's Montana Brewery Museum.)

This check from the H.S. Gilbert Brewery was made out to the Madison State Bank (the brewery was located in Madison County, Montana), and bears the signature of Valentine Gilbert. (Courtesy of Lozar's Montana Brewery Museum.)

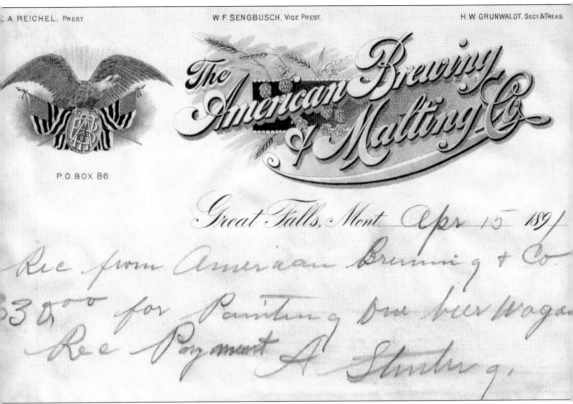

The American Brewing & Malting Co.

P.O. BOX 86.

Great Falls, Mont. Apr 15 1891

Rec from American Brewing & Co.

$30.00 for Painting one beer Wagon.

Rec Payment A. Sturber.

It is the proverbial paper trail that provides unique glimpses into history. Here is an invoice from American Brewing Company (Great Falls) dated April 15, 1891, for $30 paid to paint "one beer wagon." (Courtesy of Lozar's Montana Brewery Museum.)

CAPITAL BREWING COMPANY

HELENA, MONTANA

C. DILLMAN, President   CAPACITY 100,000 BARRELS

utte Office, 121 Pennsylvania Building Telephone, 45

Brewers
of
Lager Beer
and
High Life
Bottled Beer

Brewers
of
Lager Beer
and
High Life
Bottled Beer

$1,000.00 REWARD

To Anyone Proving That there is Used in the Manufacture of This Beer
Any Substitute For Malt and Hops

'he Best Bottled Beer in the Market - 500,000 Bottles Sold Within the Six Mon

Another paper trail is the one left in advertising. In this advertisement, Capital Brewing Company (Helena) offers its customers a $1,000 reward "to anyone proving that there is used in the manufacture of this beer any substitute for malt and hops." Additionally, they claim 500,000 bottles were sold in the last six months. (Courtesy of Lozar's Montana Brewery Museum.)

A lone relic of a small brewery, this early keg is from the Gallatin Brewing Company. (Courtesy of Lozar's Montana Brewery Museum.)

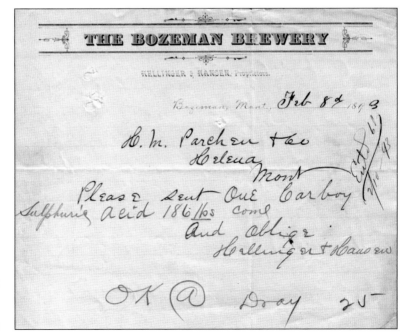

No Reclamation Allowed Unless Made Within Three Days After Receipt of Goods.

TELEPHONE 32.

Bozeman BREWERY,

JULIUS LEHRKIND,
Proprietor

Bozeman, Mont. MAY 26 1919 191

Sold to Livingston Ice. Co.

Livingston

TERMS

ALL KINDS OF CARBONATED DRINKS.

DATE	CASES	½	¼		PRICE	
				52000 # Ice	$117 00	
				Car # 37471 CB&Q.		

In Gallatin County, the Bozeman Brewery, founded by Julius Lehrkind, lasted until 1919. Julius and his brother Fred also owned breweries in Red Lodge and Silesia. In 1860, Julius stowed away on a ship at age 17 to avoid the German draft. His Red Lodge brewery produced 35,000 barrels a year and was marketed as Montana Bud. On this receipt, dated near the time the brewery closed, Lehrkind bought 52,000 pounds of ice for $117. Note that the receipt identifies the business as providing "all kinds of carbonated drinks." (Courtesy of Lozar's Montana Brewery Museum.)

In another receipt for the Bozeman Brewery, dated in the 1890s, the brewery requested barley and sulfuric acid. (Courtesy of Lozar's Montana Brewery Museum.)

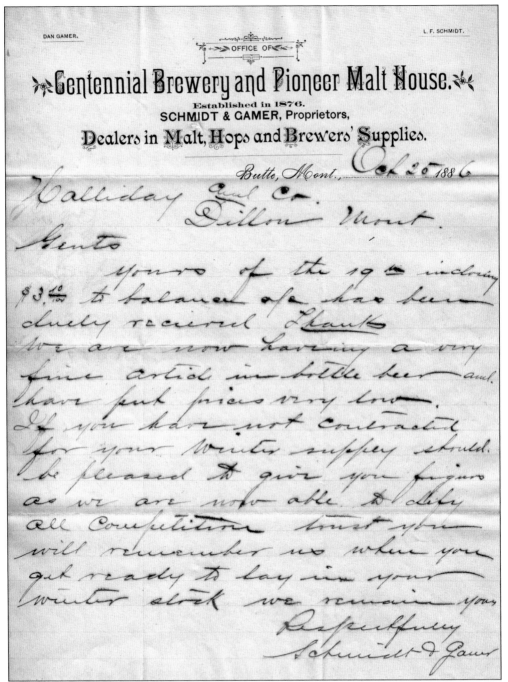

DAN GAMER.

L. F. SCHMIDT.

OFFICE OF

Centennial Brewery and Pioneer Malt House.
Established in 1876.
SCHMIDT & GAMER, Proprietors,
Dealers in Malt, Hops and Brewers' Supplies.

Butte, Mont., Oct 20 1886

Halliday and Co.

Dillon Mont.

Gents

Yours of the 19th inclosing $3.40 to balance a/c has been duly received Thanks We are now having a very fine article in bottle beer and have put prices very low. If you have not contracted for your winter supply should be pleased to give you figures as we are now able to defy all Competition trust you will remember us when you get ready to lay in your winter stock we remain yours

Respectfully

Schmidt & Gamer

This document is dated 1886 and comes from the Centennial Brewing Company (Butte) and Pioneer Malt House. It appears to clear up a balance of about $3 for some transaction. The writer goes on to state, "We are having a very fine article in bottled beer." (Courtesy of Lozar's Montana Brewery Museum.)

Drink Centennial Beer
THE BEER FOR THE PEOPLE
BEST IN BUTTE, TRY A CASE

Missoula Trust & Savings Bank

Capital - $200,000.00
$ $ $ $ $ $ $ $ $ $ $ $ $
Surplus - - $50,000.00

General Banking Business

South Butte Coal Company

C O A L
WOOD

Butte - - - - - Montana

Blue Ribbon Buffet

Peter Peterson

819 EAST FRONT STREET
South Butte - - Montana

A, 1 P. M. Special cars will run regularly ever few minutes to the Columbia Gardens, where the exercises of the day will be held.

Missoula Cab and Transfer Co.

Ind. Phone 633
Bell Phone 33

EAST MAIN STREET, Opp. Opera House

OLYMPIA BREWING CO.

"Exquisit"

BEST MONTANA PRODUCT

F. A. Gagner Fine Liquors and Cigars

On this pamphlet from Butte, Montana, nearly every facet that is connected to drinking beer can be found. The advertisements are for two breweries, Centennial and Olympia Brewing Companies, a taxi company, a bank, a buffet, and one for cigars and liquors. One can imagine how each business related to the others and shared customers. (Courtesy of Lozar's Montana Brewery Museum.)

The Billings Brewery, founded in 1900, understood marketing in a way not many other Montana breweries knew at the time. Old Fashion Beer was its flagship, and before the brewery could even open its doors to release its first batch, 3,000 people showed up for a drink. (Courtesy of Lozar's Montana Brewery Museum.)

Not only did the Billings Brewery have one of the largest electric signs in Montana at the time, a 40-by-25-foot sign fitted with 920 bulbs that lit in sequence, it also had this unique beer bottle car, which debuted in 1910. The president of the brewery, Phil Grein, drove around in it frequently. Unfortunately, the car was eventually scrapped for metal. (Courtesy of Lozar's Montana Brewery Museum.)

Billings was nicknamed "Magic City" because of its rapid growth. To add to the magic was this electric sign for Old Fashion Beer, the largest electric sign in Montana. (Courtesy K. Ross Toole Archives, the University of Montana–Missoula.)

The Lewistown Brewery had a history rich in mystery, intrigue, and late-night border crossings. It was founded in 1894 by Frank Hass and constructed of locally quarried sandstone by John Laux, the founder's brother. (Courtesy of Lozar's Montana Brewery Museum.)

After Prohibition, Lewistown Brewing Company did not stick to making near beer, as others did. Instead they bribed their local Prohibition enforcement agents to look the other way. However, in 1922, the gig was up, and one agent was indicted on 12 counts of accepting bribes. The agent received bribes from Lewistown Brewing and Montana Brewing Company in Great Falls. (Courtesy of Lozar's Montana Brewery Museum.)

After Prohibition, Lewistown Brewing tried to survive by producing 25 barrels of beer a day. These barrels were lined with pitch. One of their two beers, Silver Tip, was brewed in a 25-barrel copper kettle. Initially, the beer was only available in kegs, but soon the brewery bottled the beer in 12-ounce (label shown) and 24-ounce bottles. (Courtesy of Lozar's Montana Brewery Museum.)

Try as it might, the brewery could not come back from Prohibition. There was too much pressure from the national brands that could supply beer across state lines cheaper than the beer made in the state. Like others, Lewistown Brewing could not last long, and it closed its doors in 1938. (Courtesy of Lozar's Montana Brewery Museum.)

Unfortunately, beer could not bring everyone together over the decades. These signs were hung over a bar in Birney, Montana, in 1941. (Photograph by Marion Post Wolcott, courtesy of the Library of Congress.)

This photograph shows the Birney beer parlor in 1939. Perhaps indicative of how things are slow to change in some places, like the treatment of Native Americans, there is irony that horses were ridden to a bar and hitched by gasoline pumps. (Courtesy of the Library of Congress.)

This is another photograph of the Birney beer parlor from 1939. Clearly, the parlor was fond of serving Pabst Blue Ribbon. (Courtesy of the Library of Congress.)

This interior shot of the Birney beer parlor shows three men enjoying song and beer. Coca-Cola and Pabst Blue Ribbon are in ample supply. (Courtesy of the Library of Congress.)

This photograph was taken in October 1942 as a keg of Butte Brewing Company beer was opened at the scrap salvage campaign in Butte. (Courtesy of the Library of Congress.)

Only open for three years (1905–1908), Basin Brewing Company was one of only two breweries to ever exist in the small town of Basin, Montana. At its peak in 1905, the mining town boasted a population of 1,500, four rooming houses, a drug store, three hotels, a bath house, three grocery stores, a bank, a newspaper, and twelve saloons. Today, Basin's population is listed at 267. (Courtesy of Lozar's Montana Brewery Museum.)

The Kalispell Malting and Brewing Company launched in 1894 when Gust Gamer and two German immigrant brothers, Henry and Charles Lindlahr, opened its doors on the corner of Fifth Avenue West and Center Street in Kalispell, Montana. (Courtesy of Lozar's Montana Brewery Museum.)

GLACIER BOCH BEER
Is Here

For centuries, the spring tonic has been Boch Beer!

This brew, an invigorating beverage, is even more healthful, more refreshing, than the regular GLACIER Beer.

It is heavier, tastier. For three months it has been mellowing and aging.

PROSIT. Get your GLACIER BOCH at your favorite dealer's today.

GLACIER BOCH BEER
The Kalispell Malting and Brewing Co.
(YOUR NEIGHBORS FOR 41 YEARS)

Local boosters called Kalispell the "Gateway to Glacier Park" after the national park was created in 1910. It was fitting for the brewery to have a beer named Glacier. (Courtesy of Lozar's Montana Brewery Museum.)

In conjunction with the brewery, the Kalispell Malting and Brewing Company also ran the Brewery Saloon at 102 Main Street. It was one of Main Street's first brick buildings. Draft beer cost 5¢ a glass, and lunch was free. The saloon was even connected to the brewery via telephone. (Courtesy of Lozar's Montana Brewery Museum.)

The Brewery Saloon expanded in 1900 to include a second story, which was dubbed the Kalispell Club and was open to men only. One year later, the Lindlahr brothers added to the saloon by opening Kalispell's first bowling alley. (Courtesy of Lozar's Montana Brewery Museum.)

After Charles Lindlahr passed away in 1898, the shares of the Lindlahr brothers were purchased by Capt. Frederick Pabst, who bought them for his nephew Christ Best. Christ was a descendent of Jacob Best, who founded the Phillip Best Brewing Company in 1844. (Courtesy of Lozar's Montana Brewery Museum.)

Gus Bischoff came on board with the brewery in 1906. He would help it survive Prohibition by making soda, cider, and near beer. In 1935, Bischoff purchased the brewery and ran it with one family member or another until it closed in 1955. (Courtesy of Lozar's Montana Brewery Museum.)

A serving tray like this would have likely been used at the Brewery Saloon, either serving up the free lunch, or when Prohibition hit, smoking supplies or fountain drinks. During Prohibition, the saloon changed its name to the Palm, and part of its income came from allowing women to bowl once a week at the men-only club. (Courtesy of Lozar's Montana Brewery Museum.)

By the time the brewery closed in 1955, it was the city's oldest business. Its bottle plant workers had been earning a better-than-average wage, and the brewery would no longer buy local barley for its beers. (Courtesy of Lozar's Montana Brewery Museum.)

In this unique photograph from the Kalispell Brewing Company, it can be understood that Pres. Franklin D. Roosevelt was widely appreciated in the brewing industry since he was the president who helped Congress pass the 21st Amendment, which repealed the 18th, ending Prohibition. (Courtesy of Lozar's Montana Brewery Museum.)

Currently in Montana there are over 90 breweries in operation, and in many of those can be found links to the breweries that came decades before. Dave Ayers, founder of Glacier Brewing Company in Polson, Montana, displays many breweriana in his brewery. His sign is patterned exactly like the one found over the H.S. Gilbert Brewery in Virginia City. (Author's collection.)

For comparison, this is the sign from the H.S. Gilbert Brewery. (Courtesy of the Library of Congress.)

It was common, as seen at H.S. Gilbert Brewery, to have little between the brewery and where patrons enjoyed the beer. That trend has continued in today's breweries as well. (Courtesy of Lozar's Montana Brewery Museum.)

Bozeman Brewing Company released Bozone Select, a tribute to Great Falls Select, one of the most popular Montana-made beers in the mid-20th century. (Courtesy of the Bozeman Brewing Company.)

Harvest Moon Brewing Company in Belt, Montana, partnered with the owner of the Great Falls Select trademark to release a new version of an old classic. Though a different style, the banner image on the can is reminiscent of the original Great Falls Select logo. (Courtesy of Eagle Beverage.)

Sometimes Montana breweries are opened in buildings with historical significance. Philipsburg Brewing Company, opened in 2012, is inside the Sayrs Building, a 125-plus-year-old former bank. (Author's collection.)

This fine collection of historic Montana beer labels can be seen integrated into the bar at Glacier Brewing Company in Polson, Montana. (Author's collection.)

The Kessler Brewing Company was revived in name only during the 1990s and lasted until 2000. It released several beer styles, including Legislative lager and Elk Hunter's Brew. (Author's collection.)

At one time, Pabst Blue Ribbon, traditionally brewed in Milwaukee at the P.H. Best Brewing Company, was brewed in Troy, Montana, because it was cheaper than paying the tariff to bring the beer across state lines. (Courtesy of Wikipedia.)

Montana has a rich brewing tradition, but the bars in which its beers are served are as historically significant and diverse as the breweries themselves. In fact, at the Sip and Dip in Great Falls, patrons may even catch a glimpse of a mermaid while drinking a Montana-made beer. (Courtesy of GreatFallsGirl.)

DISCOVER THOUSANDS OF LOCAL HISTORY BOOKS FEATURING MILLIONS OF VINTAGE IMAGES

Arcadia Publishing, the leading local history publisher in the United States, is committed to making history accessible and meaningful through publishing books that celebrate and preserve the heritage of America's people and places.

Find more books like this at
www.arcadiapublishing.com

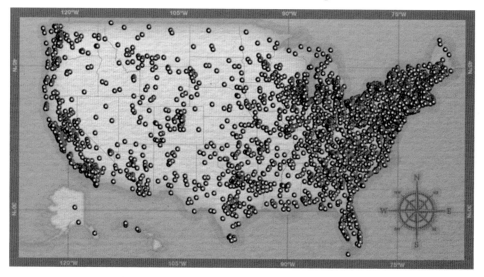

Search for your hometown history, your old stomping grounds, and even your favorite sports team.

Consistent with our mission to preserve history on a local level, this book was printed in South Carolina on American-made paper and manufactured entirely in the United States. Products carrying the accredited Forest Stewardship Council (FSC) label are printed on 100 percent FSC-certified paper.

MADE IN THE
USA